EARTH SCIENCE—GEOLOGY Need to Know

SilverTip

Metamorphic Rocks

by Ruth Owen

Consultant: E. Calvin Alexander Jr., Professor Emeritus
Earth & Environmental Sciences
University of Minnesota, Minneapolis

BEARPORT PUBLISHING

Minneapolis, Minnesota

Credits
Cover and Title Page, © BlackRabbit3/Shutterstock; 3, © Nikitin Victor/Shutterstock; 5T, © Iurii Osadchi/Shutterstock; 5M, © Alhim/Shutterstock; 5B, © EzumeImages/iStock; 6, © PTZ Pictures/ Shutterstock; 7, © Jacob Boomsma/Shutterstock; 9, © Nektarstock/Shutterstock; 11Yuri_ArcursiStock; 12, © Shutterstock; 13, © BlackRabbit3/Shutterstock; 14, © Designua/Shutterstock; 15, © Designua/ Shutterstock; 17, © Richard Bradford/Shutterstock; 19 (from top), © EVGEIIA/Shutterstock, © Koenig_K/ Shutterstock, © gingerm/Shutterstock, © Trygve Finkelsen/Shutterstock; 21, © LesPalenik/Shutterstock; 21 (inset), © Fokin Oleg/Shutterstock; 23, © Alessandro Colle/Shutterstock; 24, © Daria Volyanskaya/ Shutterstock; 25T, © Bjoern Wylezich/Shutterstock; 25B, © IanDagnall Computing/Alamy; 27, © Roman Khomlyak/Shutterstock; 28, © VectorMine/Shutterstock.

President: Jen Jenson
Director of Product Development: Spencer Brinker
Senior Editor: Allison Juda
Associate Editor: Charly Haley
Designer: Colin O'Dea

Library of Congress Cataloging-in-Publication Data

Names: Owen, Ruth, 1967- author.
Title: Metamorphic rocks / Ruth Owen.
Description: Silvertip books. | Minneapolis, Minnesota : Bearport Publishing Company, [2022] | Series: Earth science - geology: need to know | Includes bibliographical references and index.
Identifiers: LCCN 2021026710 (print) | LCCN 2021026711 (ebook) | ISBN 9781636915784 (library binding) | ISBN 9781636915852 (paperback) | ISBN 9781636915920 (ebook)
Subjects: LCSH: Metamorphic rocks–Juvenile literature.
Classification: LCC QE475.A2 O94 2022 (print) | LCC QE475.A2 (ebook) | DDC 552/.4–dc23
LC record available at https://lccn.loc.gov/2021026710
LC ebook record available at https://lccn.loc.gov/2021026711

For more information, write to Bearport Publishing, 5357 Penn Avenue South, Minneapolis, MN 55419.
Printed in the United States of America.

Contents

Rock Our World

What do pool tables, shower tiles, and pizza stones have in common? They can all be made with **metamorphic** (*met*-uh-MOR-fik) rocks! There are many different metamorphic rocks. But they all form in the same ways underground.

There are three kinds of rocks. They are metamorphic, **igneous** (IG-nee-uhs), and **sedimentary** (*sed*-uh-MEN-tur-ee). Rocks are put in these groups based on the different ways they form.

There is metamorphic rock under the fabric of a pool table.

What Are Rocks Made Of?

Almost all rocks are made of **minerals**. Minerals are made by nature. Some rocks have just one mineral. Others have several different minerals. And rocks look different based on which minerals are in them.

The metamorphic rock quartzite is made of the mineral quartz.

Minerals give rocks their colors. Rocks with more than one mineral may have patterns with different colors. Some have stripes or spots.

All Change!

How do metamorphic rocks form? The clue is in their name! The name comes from a Greek word that means change in form. A metamorphic rock can start as any kind of rock. This rock is called the parent rock. The parent rock goes through a change to become metamorphic.

Igneous and sedimentary rocks can become metamorphic rocks. Metamorphic rocks can also change into other metamorphic rocks.

Baking New Rocks

Metamorphic rocks change when there is high heat or **pressure**. Sometimes, there is both. This change happens deep underground. Earth is covered in a **crust** of rock. Below the crust is hot, melted rock called **magma**.

Earth's crust is about 19 miles (30 km) deep where there is land. It is only about 3 miles (5 km) deep beneath oceans.

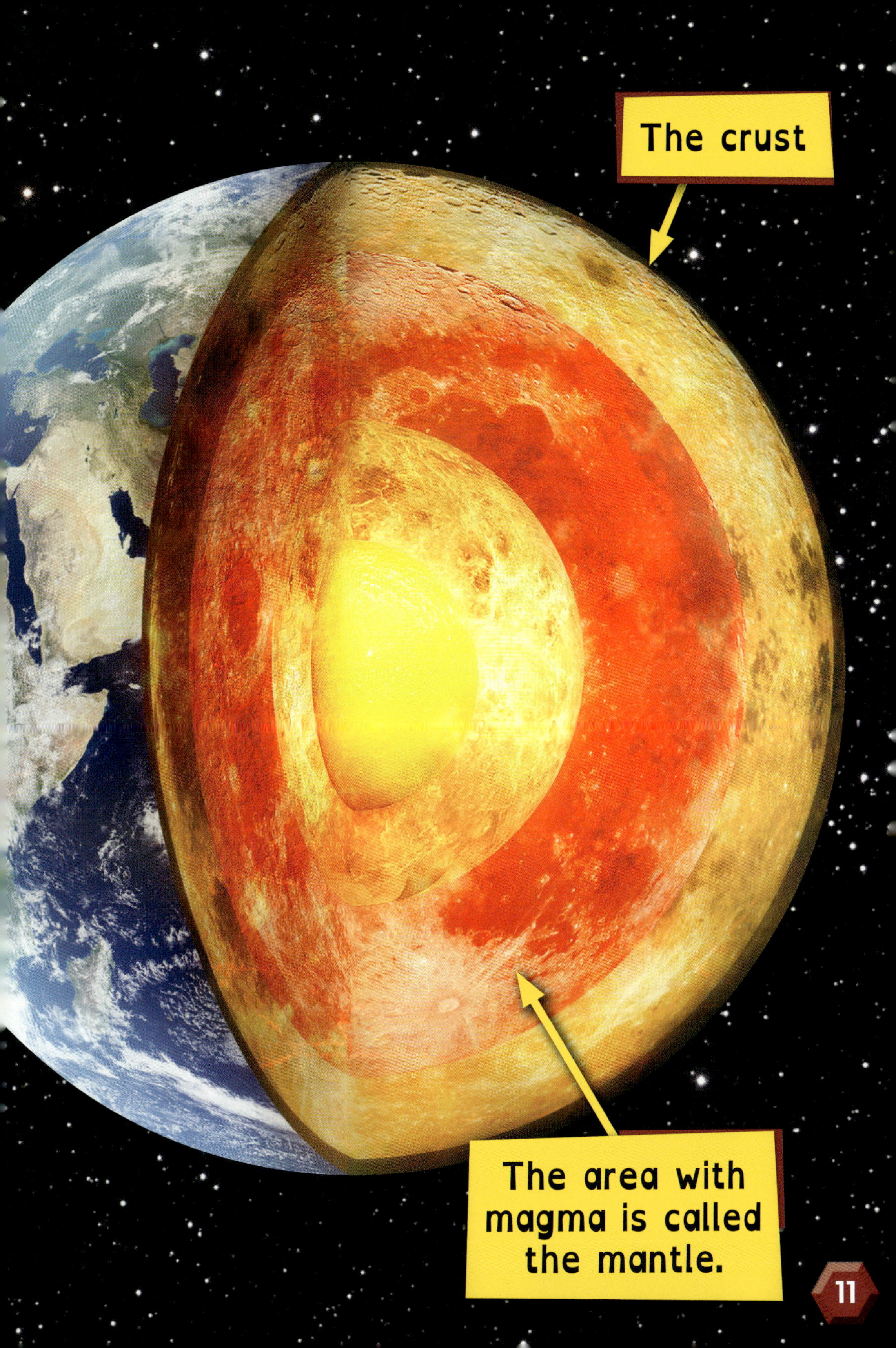
The crust
The area with magma is called the mantle.

Sometimes, magma flows into cracks in the crust and heats the rock around it. This changes the rock's minerals into something new. They become metamorphic rock! It is a little like cookie dough in the oven. The ingredients of the rock stay the same. The heat just forms them into something new.

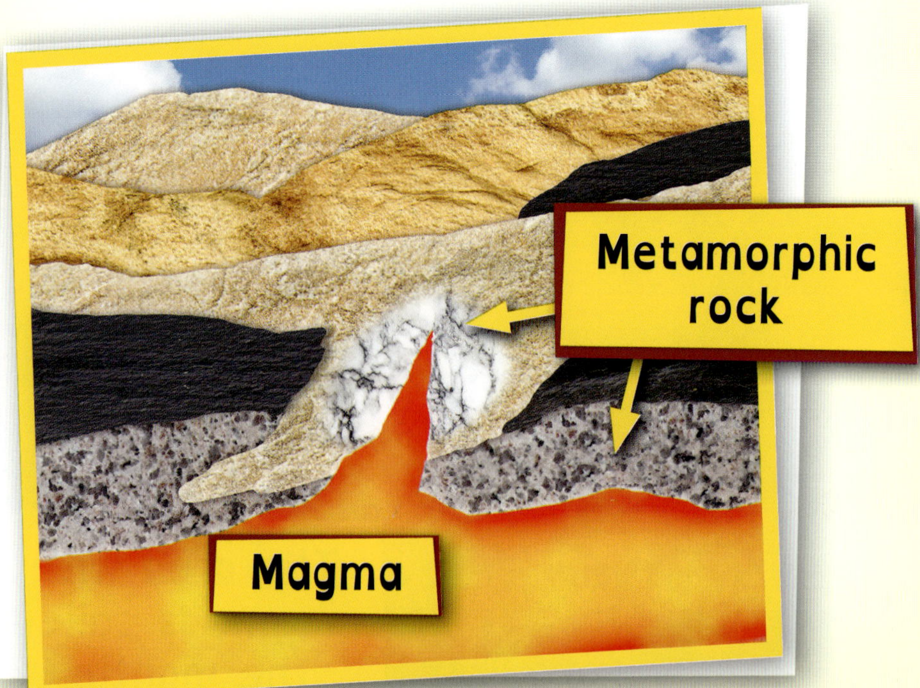

A metamorphic rock called hornfels can form when sedimentary shale and sandstone are baked. The rocks must be heated to about 1,450 degrees Fahrenheit (788 degrees Celsius) to make this change.

Rocky Plates

Baking from magma is not the only way metamorphic rocks form in Earth's crust. The crust is made of giant pieces called **tectonic plates**. They fit together like a puzzle. The places where these plates meet are called boundaries. The plates are always slowly moving.

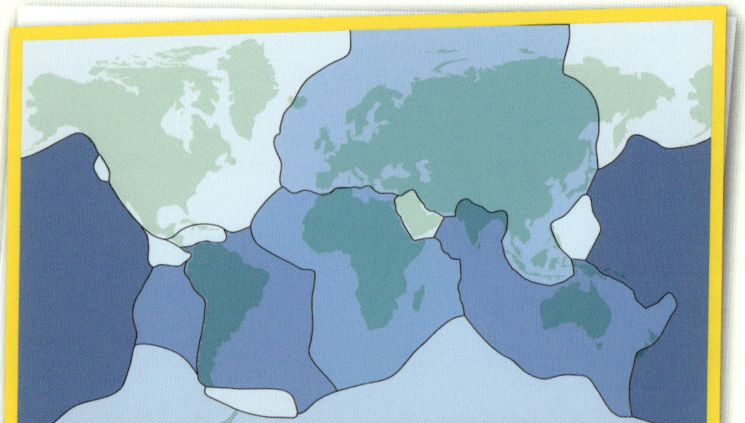

There are seven major tectonic plates and many smaller ones.

Tectonic plates are very slow. They move only a few inches a year. As they move, they can rub together. Sometimes, they crash into one another or pull apart.

The moving plates create a lot of pressure and heat. The rocks at the boundaries get baked. They change into metamorphic rock.

When tectonic plates move, they push metamorphic rocks deeper into Earth's crust. This causes even more pressure and heat. The rocks get hotter as they are pushed deeper.

The metamorphic rock gneiss is part of this cliff.

Changing Again and Again

With more heat and pressure, metamorphic rocks keep changing. Over millions of years, a rock may change again and again into several kinds of metamorphic rock!

Eventually, a rock that is very deep in Earth's crust can be crushed and heated so much that it melts! Then, it is no longer metamorphic rock. It has become magma.

The metamorphic rock slate can turn into phyllite. If it keeps changing, it can become schist and then gneiss.

Striped Rocks

The way metamorphic rocks form can create stripes in the rock. How? The different minerals in the parent rock may be mixed up. But then, the rock goes through pressure and heat. As the rock changes, its minerals group together again.

Some metamorphic rocks have swirled patterns. The swirls are parts of the rock that started melting but then hardened again.

The igneous rock granite has minerals mixed up.

When granite turns into metamorphic gneiss, its minerals line up in stripes.

Coming to the Surface

How can we see these rocks that form and change underground? Wind, rain, and snow wear away at Earth's surface. This breaks down the top **layers** of rock. Then, we see the metamorphic rock that was once underground. Sometimes, movements in Earth's crust can push rock up to the surface.

People also dig to get metamorphic rock. Mountains in Italy hold the metamorphic rock marble. People have cut marble from the mountains for more than 2,000 years.

A Beautiful Blue Rock

People have used metamorphic rock in many ways throughout history. Lapis lazuli (LA-eh-pis la-eh-ZOO-lee) is a beautiful blue metamorphic rock. It has been used for about 9,000 years. The ancient Egyptians used it to make beads, rings, and small statues.

People have also crushed lapis lazuli to make blue paint. The artist Vincent van Gogh used it in his famous painting *The Starry Night*.

The Starry Night

Always Changing

It may take 20 to 30 million years for metamorphic rock to form in Earth's crust. Then, it can take millions of years for the rock to come to the surface. But these changes never stop. Metamorphic rocks are always forming and changing. New rock is forming deep below your feet right now!

Death Valley National Park has a **canyon** with metamorphic marble. The marble formed underground. But then tectonic plates shifted. Water cut through the rocks and made the canyon. Now, the marble can be seen.

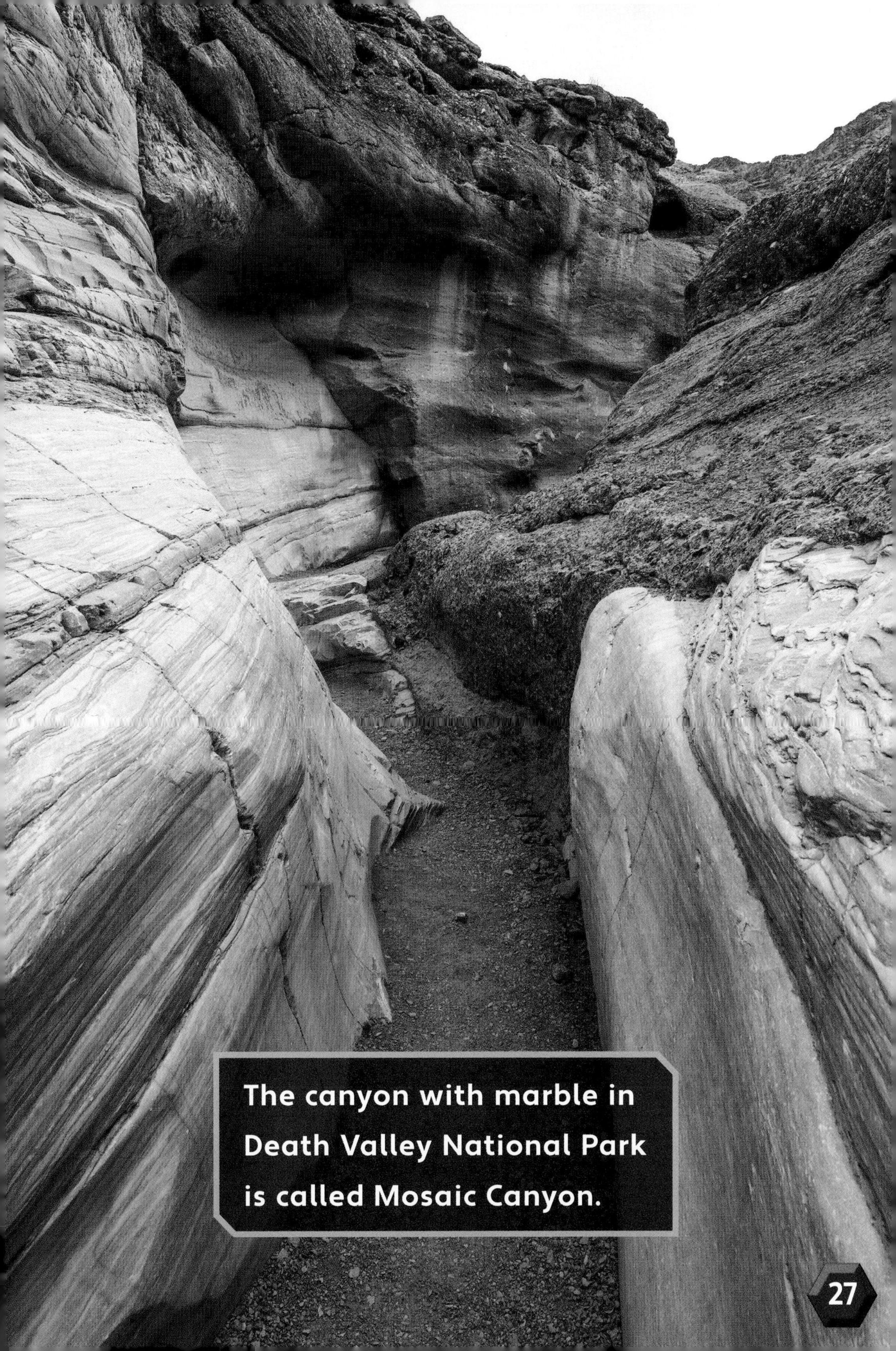

The canyon with marble in Death Valley National Park is called Mosaic Canyon.

The Changes of Metamorphic Rock

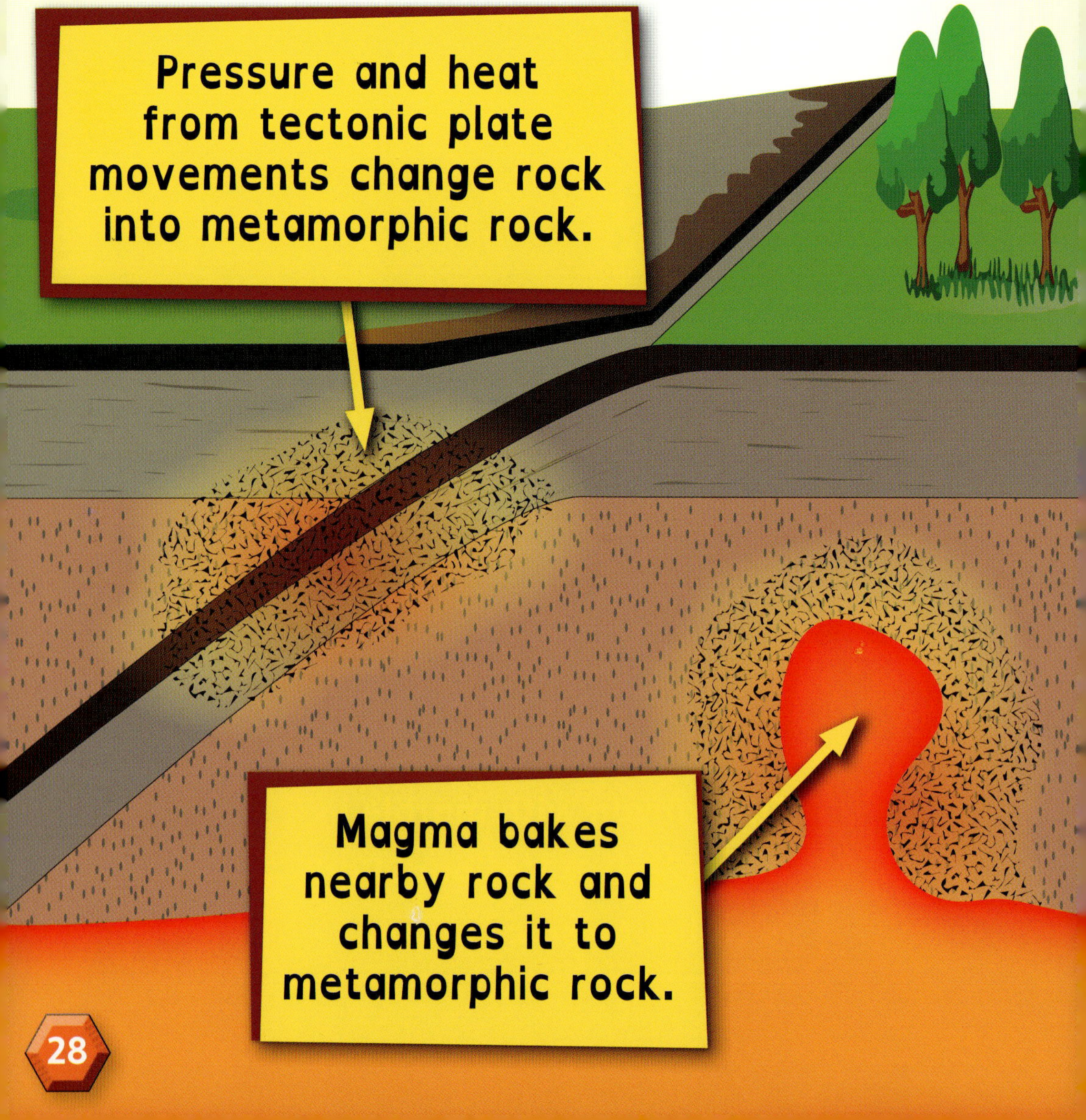

SilverTips for SUCCESS

★ SilverTips for REVIEW

Review what you've learned. Use the text to help you.

Define key terms

boundaries
crust
magma
minerals
tectonic plates

Check for understanding

What are the three different kinds of rocks?

Describe two ways metamorphic rocks are made. What is the same about these ways and what is different?

What role can heat play in forming metamorphic rocks?

Think deeper

How do people use metamorphic rocks? Why do these rocks matter?

★ SilverTips on TEST-TAKING

- **Make a study plan.** Ask your teacher what the test is going to cover. Then, set aside time to study a little bit every day.
- **Read all the questions carefully.** Be sure you know what is being asked.
- **Skip any questions** you don't know how to answer right away. Mark them and come back later if you have time.

Glossary

canyon a steep valley carved out by a river

crust the outer layer of Earth, made of rock

igneous one of the three main types of rock that forms from magma or lava that has cooled and become solid

layers levels of something, such as rocks, lying on top of or under another level

magma hot liquid rock flowing inside Earth

metamorphic one of the three main types of rock that forms when rock changes from one type to another because of heat or pressure

minerals solid substances found in nature that make up rocks

pressure the force produced when something presses on something else

sedimentary one of the three main types of rock that forms from tiny pieces of rock

tectonic plates sheets of rock that make up Earth's crust

Read More

Fretland VanVoorst, Jenny. *Metamorphic Rocks (Blastoff! Discovery: Rocks & Minerals).* Minneapolis: Bellwether Media, 2020.

Pettiford, Rebecca. *Metamorphic Rocks (Geology Genius).* Minneapolis: Jump! 2019.

Rogers, Marie. *Exploring Metamorphic Rocks (Let's Rock!).* New York: PowerKids Press, 2021.

Learn More Online

1. Go to **www.factsurfer.com** or scan the QR code below.
2. Enter "**Metamorphic Rock**" into the search box.
3. Click on the cover of this book to see a list of websites.

Index

About the Author

Ruth Owen has written hundreds of non-fiction books. She lives on the rocky Cornish coast in England and has always been fascinated by rocks.